CHANDRAYAAN-3 MISSION:

A moment of glory for India

MATTHEW ESSIE

Copyright claim
All rights reserved. No part of this publication may be reproduced, distributed, or by any means, including photocopying, recording, or other electronic or mechanical methods, without the prior written permission of the publisher, except in the case of brief quotations embodied in critical reviews and certain other noncommercial uses permitted by copyright law

TABLE OF CONTENT

CHAPTER 1

WHAT IS CHANDRAYAAN 3

The third and most recent lunar exploration mission conducted by Indian Space Research is called Chandrayaan-3. It is similar to Chandrayaan-2 in that it comprises of a lander named Vikram and a rover named Pragyan, but it lacks an orbiter. Its propulsion system functions like a satellite relaying communications. Up until the spacecraft is in a 100 km lunar orbit, the lander and rover combination is carried by the propulsion module.

With the primary goal of deploying a lander and rover in the highlands near the south pole of the Moon in late August 2023 and showcasing end-to-end landing and roving capabilities, Chandrayaan 3 is an ISRO (Indian Space Research Organization) project.

Additionally, it will perform a variety of scientific measures both from orbit and on the ground. It consists of a propulsion module and a lander/rover. Similar to the Vikram rover on Chandrayaan 2, the lander/rover will include upgrades to aid guarantee a secure landing. The propulsion module, which will remain in lunar orbit and serve as a communications relay satellite, will lift it into lunar orbit.

To examine the spectral and Polarimetric data of Earth from the lunar orbit, the propulsion module is equipped with the Spectro-Polarimetry of Habitable Planet Earth (SHAPE) payload.

Lander payloads include the Langmuir Probe (LP), Chandra's Surface Thermophysical Experiment (ChaSTE), Instrument for Lunar Seismic Activity (ILSA), and Chandra's Surface Thermophysical Experiment (ChaSTE), which measures thermal conductivity and temperature. For lunar laser ranging investigations, the space agency NASA has provided a passive Laser Retroreflector Array.

Alpha Particle X-ray Spectrometer (APXS) and Laser Induced Breakdown Spectroscope (LIBS) are rover payloads that are used to determine the elemental composition close to the landing site.

To research and showcase new technology necessary for interplanetary missions, Chandrayaan-3 is made up of an indigenous Lander module (LM), Propulsion module (PM), and Rover. The Lander will be able to gently land at a chosen location on the moon and release the Rover, which will conduct an in-situ chemical analysis of the lunar surface as it is moving. There are scientific payloads on the Lander and the Rover that will conduct lunar surface tests. The main job of PM is to transport the LM from injection into the launch vehicle to the final 100 km circular polar orbit of the moon and then to release the LM from PM.

In addition to this, the Propulsion Module features one additional scientific payload that will be operated post separation of the Lander Module. The launcher identified for Chandrayaan-3 is LVM3 M4 which will place the integrated module in an Elliptic Parking Orbit (EPO) of size ~170 x 36500 km.

CHAPTER 2

PROPERTIES OF CHANDRAYAAN-3

S/N	PARAMETERS	SPECIFICATIONS
1	Mission Life (Lander & Rover)	One lunar day (~14 Earth days)
2	Landing Site (Prime)	4 km x 2.4 km 69.367621 S, 32.348126 E
3	Science Payloads	Lander: • Radio Anatomy of Moon Bound Hypersensitive ionosphere and Atmosphere

		(RAMBHA) • Chandra's Surface Thermo physical Experiment (ChaSTE) • Instrument for Lunar Seismic Activity (ILSA) • Laser Retroreflector Array (LRA) Rover: • Alpha Particle X-Ray Spectrometer (APXS) • Laser Induced Breakdown Spectroscope (LIBS) Propulsion Module: • Spectro-polarimetry of HAbitable Planet Earth (SHAPE)
4	Two Module Configuration	• Propulsion Module (Carries Lander from launch injection to Lunar orbit)

		• Lander Module (Rover is accommodated inside the Lander)
5	Mass	☐ **Propulsion Module**: 2148 kg ☐ **Lander Module**: 1752 kg including Rover of 26 kg Total: 3900 kg
6	Power generation	**Propulsion Module**: 758 W **Lander Module**: 738W, WS with Bias Rover: 50W
7	Communication	**Propulsion Module:** Communicates with IDSN **Lander Module:** Communicates with IDSN and Rover. Chandrayaan-2 Orbiter is also planned for contingency link.

		Rover: Communicates only with Lander.
8	Lander Sensors	<ul><li>Laser Inertial Referencing and Accelerometer Package (LIRAP)</li><li>Ka-Band Altimeter (KaRA)</li><li>Lander Position Detection Camera (LPDC)</li><li>LHDAC (Lander Hazard Detection & Avoidance Camera)</li><li>Laser Altimeter (LASA)</li><li>Laser Doppler Velocimeter (LDV)</li><li>Lander Horizontal Velocity Camera (LHVC)</li><li>Micro Star sensor</li><li>Inclinometer & Touchdown sensors</li></ul>
9	Lander Actuators	Reaction wheels – 4 nos (10

		Nms & 0.1 Nm)
10	Lander Propulsion System	Bi-Propellant Propulsion System (MMH + MON3), 4 nos. of 800 N Throttleable engines & 8 nos. of 58 N; Throttleable Engine Control Electronics
11	Lander Mechanisms	<ul><li>Lander leg</li><li>Rover Ramp (Primary & Secondary)</li><li>Rover</li><li>ILSA, Rambha & Chaste Payloads</li><li>Umbilical connector Protection Mechanism,</li><li>X- Band Antenna</li></ul>
12	Lander Touchdown specifications	**Vertical velocity:** ≤ 2 m / sec **Horizontal velocity:** ≤ 0.5 m / sec **Slope:** ≤ 12 deg

What distinguishes the Chandrayaan-3 Mission from its predecessor

The difference between Chandrayaan-3 and its failed predecessor is that SHAPE, which stands for Spectro-polarimetry of HAbitable Planet Earth, is a payload aboard the propulsion module that will be used to examine Earth from lunar orbit.

According to the ISRO, the SHAPE is an experimental payload designed to investigate the Earth's spectro-polarimetric fingerprints in the region of near-infrared wavelengths.

The Propulsion Module's primary duty, in addition to carrying the SHAPE payload, is to transport the Lander Module from the launch vehicle's injection orbit until lander separation. After touching down on the moon's surface, the Lander Module carried many payloads, including RAMBHA-LP, which was designed to assess variations in the density of plasma ions and electrons near the surface.

The ChaSTE Chandra's Surface Thermo Physical Experiment and the ILSA (Instrument for Lunar Seismic Activity) instruments will study the thermal characteristics of the lunar surface close to the poles and map the crustal and mantle structures of the moon, respectively. After soft-landing, the rover would exit the Lander Module and use its payloads to observe the moon's surface. Alpha Particle X-Ray Spectrometer (APXS)

CHAPTER 3

HISTORY AND LATEST NEWS ON CHANDRAYAAN-3

History

With the successful launch of its Chandrayaan-3 mission on Friday, India is attempting to become just the fourth nation to successfully carry out a controlled landing on the moon.

Soon after 2:30 p.m. local time (5 a.m. ET), Chandrayaan—which is Sanskrit for "moon vehicle"—launched from the Satish Dhawan Space Center in Sriharikota, southern Andhra Pradesh.

More than 1 million people watched the historic launch online in addition to the large crowds that assembled at the space complex.

Later on Friday, the Indian Space Research Organization announced through Twitter that Chandrayaan-3 is in "precise orbit" and has "begun its journey to the moon."

The health of the spacecraft was further described as "normal."

A new chapter in India's space voyage is scripted by Chandrayaan-3, the Indian Prime Minister Narendra Modi tweeted in response. It flies high, boosting the aspirations and desires of every Indian. This outstanding accomplishment is evidence of our scientists' unwavering commitment. I applaud their creativity and energy.

On August 23, the craft is anticipated to touch down on the moon.

India is attempting a soft landing for the second time after Chandrayaan-2, its last attempt, failed in 2019. Chandrayaan-1, its first lunar mission, orbited the moon before being purposefully crash-landed into the lunar surface in 2008.

Only the United States, Russia, and China have successfully completed the challenging task of soft-landing a spacecraft on the moon's surface.

On the launch, Indian engineers have been working for many years. They intend to set down Chandrayaan-3 close to the difficult terrain of the undiscovered South Pole of the moon.

Chandrayaan-1, India's first lunar mission, found water molecules on the moon's surface. The Chandrayaan-2 successfully entered lunar orbit eleven years later, but its rover impacted the moon's surface. It was also planned to investigate the South Pole of the moon.

Despite the mission's failure, Indian Prime Minister Narendra Modi praised the engineers behind it and pledged to continue working on India's space program and goals.

Latest news

On July 14, the orbiter, lander, and rover spacecraft were launched. On August 23 or 24, it will attempt to land the lander and rover on the lunar surface.

If the mission is successful, India will be the first nation to set foot close to the Moon's little-visited south pole.

After the US, the former Soviet Union, and China, it will be the only other country to successfully do a soft landing on the Moon.

The spacecraft was sent into the translunar orbit on Tuesday with a slingshot maneuver after orbiting the Earth for more than a week.

Chandrayaan-3, the third mission in India's lunar exploration program, is anticipated to build on the accomplishments of its predecessors.

Ancient India Moon mission successfully takes off.

It was launched 13 years after the nation's initial Moon mission in 2008, which showed that the Moon has an atmosphere during the daylight and that there are water molecules on its dry surface. The July 2019 launch of Chandrayaan-2, which also included an orbiter, a lander, and a rover, was only partially successful. Even now, the orbiter is still circling and studying the Moon, but the lander-rover crashed upon touchdown since it was unable to achieve a gentle landing.

The Chandrayaan-3 will launch, orbit the Earth in stages until it reaches the Moon's orbit, at which point the lander will separate from the propulsion module and land close to the Moon's south pole

White space for presentations

Sreedhara Panicker Somanath, the head of the Indian Space Research Organization (Isro), said they had thoroughly examined the data from the disaster and conducted modeling exercises to rectify the flaws in Chandrayaan-3, which weighs 3,900kg and cost 6.1bn rupees ($75m; £58m).

The 26kg rover, dubbed Pragyaan, which is Sanskrit for "wisdom," is carried inside the lander, which is named Vikram after the founder of Isro and weighs roughly 1,500kg.

Isro tweeted on Tuesday that the spacecraft had finished orbiting the Earth and was moving toward the Moon.

The spacecraft was successfully launched into a translunar orbit using an Isro perigee-firing. The Moon is the next stop, it stated. The part of an orbit that is closest to Earth is called a perigee.

The Moon mission launched by India—was it successful?

The female researchers who propelled India into space

Scientists will start progressively lowering the rocket's speed now that the ship has reached the Moon's orbit in order to bring it to a location that would enable a gentle landing for Vikram.

The six-wheeled rover will land, eject, and wander across the Moon's surface collecting vital information and photographs that will be sent back to Earth for examination.

"The rover is carrying five sensors that are focused on learning more about the tectonic activity under the Moon's surface, the atmosphere near the surface, and the physical properties of the Moon's surface. I'm hoping we discover something fresh" Mr Somanath has said.

The south pole of the Moon is still mostly unexplored, and since it has a bigger surface area that is always under shadow than the north pole, scientists believe that there may be water present there.

There is increasing interest in the Moon on a worldwide scale, not only in India. Additionally, experts claim there is still much to learn about the Moon, which is sometimes referred to as a portal to deep space.

Chandrayaan-3 Mission timeline

On 6 July, ISRO said that Mission Chandrayaan-3 will be launched on 14 July from the second launch pas, Sriharikota.

On 7 July, Vehicle electrical tests completed

On 11 July, the 'Launch Rehearsal' simulating the entire launch operation and process lasting 24 hours concluded.

On 14 July, LVM3 M4 vehicle successfuuly launched Chandrayaan-3 into orbit

On 15 July, the first orbit-raising maneuver (Earthbound firing-1) was successfully performed at ISTRAC/ISRO, Bengaluru. Spacecraft was 41762 km x 173 km orbit.

On 17 July, The second orbit-raising maneuver was performed. The Chandrayaan-3 spacecraft was in 41603 km x 226 km orbit.

On 22 July, the fourth orbit-raising maneuver (Earth-bound perigee firing) was completed. The spacecraft was in a 71351 km x 233 km orbit.

On 25 July, an Orbit-raising maneuver performed

On 1 August, Chandrayaan-3 was inserted into the translunar orbit. The orbit achieved was 288 km x 369328 km

On 5 August, Chandrayyan-3 was successfully inserted into the lunar orbit. The orbit achieved is 164 km x 18074 km, as intended.

CHAPTER 4

IMPORTANCE OF THE CHANDRAYAAN-3

ISRO has set three main objectives for the Chandrayaan-3 mission, which include:

Getting a lander to land safely and softly on the surface of the Moon.

Observing and demonstrating the rover's loitering capabilities on the Moon.

In-site observation & conducting experiments on the materials available on the lunar surface to better understand the composition of the Moon.

Chandrayaan-3 Mission: How important is this for India

India would join the US, China, and the erstwhile Soviet Union as the fourth nation to perfect the technique of soft-landing on the lunar surface if Chandrayaan-3 is successful in successfully deploying a robotic lunar rover.

The second attempt by ISRO in four years is Chandrayaan-3. 2008 saw the first mission of Chandrayaan. On September 7, 2019, the lander "Vikram" of Chandrayaan-2, the second mooncraft mission, crashed onto the moon's surface while attempting a gentle landing due to problems with the lander's braking system.

WHAT EXACTLY IS CHANDRAYAAN-3 RELEVANCE TO THE WORLD

Chandrayaan 3's main goal is to perform India's first soft landing on another celestial body, such as the Moon, without causing any harm. On the other side though,

it will also demonstrate new technologies required for interplanetary spaceflight and provide valuable insights into the lunar surface by exploring the shadowed regions and craters of the moon.

Three primary goals have been set for the Chandrayaan-3 mission by ISRO. These goals include accomplishing a safe and gentle landing on the Moon, displaying the rover's mobility on the lunar surface, and performing on-site scientific studies.

The aim is to explore both sides of the Moon: the visible side and the far side or the dark side.

India's Chandrayaan-3 lunar mission will be crucial to the world's future space missions, hence its success will be crucial for both India and the rest of the world. Chandrayaan 3 will primarily target the southern pole of the Moon since it has more shadowed area than the northern pole. In essence, the world will go to the Moon's coldest region thanks to India's most innovative technology.

According to scientists, certain areas on the lunar surface may have a permanent supply of water.

Also, scientists are fascinated by the craters discovered at the southern pole. They believe these cold traps may include complex fossil remains of the early planetary system.

It could also help promote technological innovation, create global alliances, and inspire the next generation of explorers and scientists.

CHAPTER 5

QUESTIONS ON CHANDRAYAAN-3

How will Chandrayaan-3 get to the lunar surface?

Both the propulsion and the Lander-Rover modules make up the spaceship. The lander-rover cargo is transported to the moon by the propulsion module. The lander-rover payload can be compared to a truck and the propulsion module to the freight.

The lander-rover payload will separate from the propulsion module and fall onto the moon once it has approached the moon's surface. The lander's engines will reduce the fall so that it approaches the moon gradually rather than crashing into it.

The rover is a small, wheeled object in the style of a trolley. The rover will exit the lander's belly once it has touched down on the moon and will then scramble over its surface.

Both the lander and the rover carry equipment for doing studies including analyzing the soil on the moon, measuring how heat is conducted through the moon's surface, and tracking the path of earthquake waves over the moon's surface.

What will Chandrayaan-3 do on the Moon?

A successful touchdown would be a major accomplishment for ISRO and would add them to a select group of countries that have successfully landed spacecraft on other planets. Beyond this accomplishment, Chandrayaan-3 has technological and scientific tasks to complete.

The Chandrayaan-3 lander's side panel will quickly unfold after landing to act as a ramp for the rover. After driving down the ramp and emerging from the lander's belly, the rover will start investigating the lunar surface.

About two weeks will pass as the solar-powered lander and rover explore their surroundings. They are not made to withstand the brisk lunar night. Only the lander, which is in direct contact with Earth, may be reached by the rover. According to ISRO, the Chandrayaan-2 orbiter can serve as a backup communications relay.

Before reaching the moon, the Chandrayaan-3 does several orbits around the earth. Once there, it makes numerous orbits around the moon before the lander separates from the propulsion module and falls to the moon's surface. Why take such an odd route?

The imaginary line that links a planet and its satellite sweeps equal regions in equal amounts of time, according to Kepler's second rule of planetary motion. This implies that the satellite moves in an elliptical orbit, speeding up as it approaches the Earth and slowing down as it departs. The rule also states that an object's velocity increases as it gets closer to the planet in proportion to how far away it is from the planet. In order to provide Chandrayaan-3 enough velocity to launch toward the moon, we want to make use of this feature.

Therefore, Chandrayaan-3 will begin orbiting the globe independently in an elliptical pattern when the LVM-3 raises it above the planet. Engineers on the ground will give it a small push when it reaches the farthest point to slightly alter its course so that its subsequent circle is larger than the previous one. As a result, the spaceship will pick up more speed when it approaches the planet during its second round. Engineers will once more slightly alter the direction when it reaches the farthest point, known as apogee, in order to provide the spaceship an even greater velocity during the third round. The spaceship will have enough velocity to launch itself toward the moon after completing 5–6 of these loops.

Once it reaches the moon, the reverse will happen. Loop-by-loop the spacecraft will get closer to the moon. When it is about 100 km from the moon's surface, the lander will detach itself and begin its descent onto the moon.

There is no parachute in our moon missions, unlike the Mars-bound probes Curiosity and Perseverance that carefully descended through the atmosphere. Isn't it easier and less expensive to parachute down than to use motors to delay the descent?

This is so because, unlike the moon, Mars has an atmosphere. The atmosphere on Mars is indeed thin. About 1% of Earth's atmospheric pressure is the norm. However, there is still an atmosphere, which incidentally is composed of carbon dioxide. To create "drag," some air must be placed underneath the parachute. The moon has none, but Mars has some.

Is the way information is sent similar to how radio stations, for example, transmit a running commentary?

No. Audio waves are used for broadcasts, and for them to travel, they need a medium like the air. Electromagnetic waves, which are energy progressions like radio waves or microwaves, are used to transmit signals via space. They can move about without a medium.